Lab Manual

by
Charles J. LaRue

AGS Publishing
Circle Pines, MN 55014-1796
800-328-2560

© 2004 AGS Publishing
4201 Woodland Road
Circle Pines, MN 55014-1796
800-328-2560 • www.agsnet.com

AGS Publishing is a trademark and trade name of American Guidance Service, Inc.

Printed in the United States of America

ISBN 0-7854-3621-9

Product Number 93911

A 0 9 8 7 6 5 4 3

Table of Contents

Safety in the Biology Laboratory

General Rules

1. Keep area clean and free of unneeded objects.

2. Do not play in the laboratory.

3. Keep equipment in good working order.

4. Wipe up spills immediately.

Heat

1. Be aware of all open flames. Flames are difficult to see on Bunsen burners.

2. Never touch flames.

3. Never touch the surface of the hotplate.

4. When heating chemicals, point test tubes away from people.

Electricity

1. Never use household current without proper supervision.

2. Remember that electricity flowing in wires causes the wires to become hot. Do not burn yourself.

Chemicals

1. Never smell chemicals directly.

2. Never taste unknown chemicals.

3. Remember that even the simplest of chemicals can cause hazardous reactions.

4. Always do chemical tests under adult supervision.

5. Do not place your fingers in your mouth while working with chemicals.

6. Always wash your hands after working with chemicals.

Microscopes

1. Carry the microscope with two hands—one hand on the arm of the microscope and the other hand beneath the base.

2. Review with your teacher the correct way to use the microscope.

3. Never use direct sunlight as the light source for the microscope.

4. Handle glass slides carefully. They break easily.

Organisms

1. Never open a container in which a mold or other organism is growing.

2. If you are allergic to certain molds, plants, or animals, tell your teacher before doing activities involving these organisms.

3. Never perform experiments that cause pain or harm to mammals, birds, reptiles, fish, or amphibians.

4. Wash your hands after handling any organism.

Some Common Biological Equipment

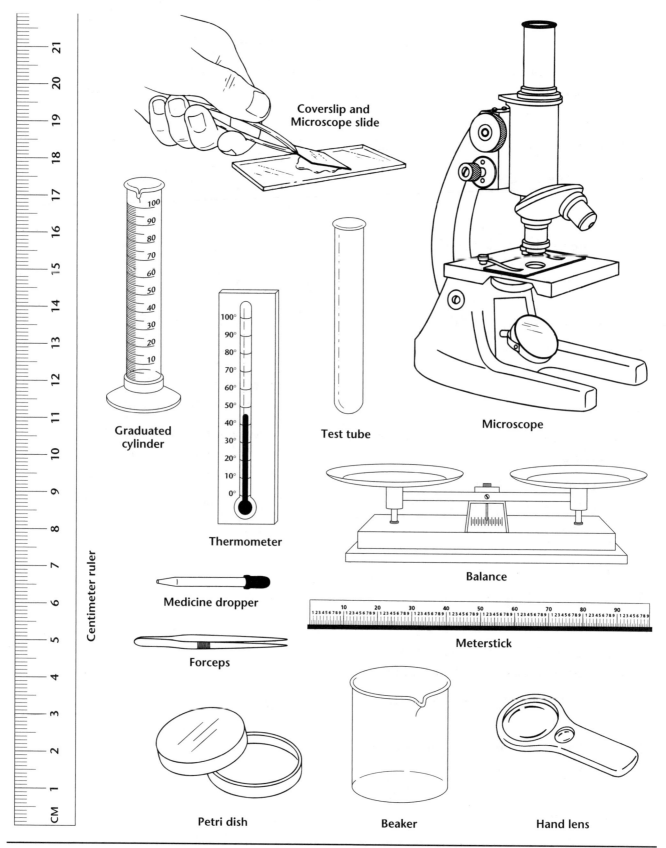

Coverslip and
Microscope slide

Graduated
cylinder

Thermometer

Test tube

Microscope

Centimeter ruler

Medicine dropper

Forceps

Balance

Meterstick

Petri dish

Beaker

Hand lens

Safety in the Classroom

Purpose: What safety equipment do you need in your classroom?
In this investigation, you will identify and locate safety equipment in
the classroom.

Materials   meterstick   metric ruler   colored pencil

Procedure

1. With a partner, draw a floor diagram of your classroom. Use the grid on the next page to make your diagram. Draw your diagram as close to the actual scale as possible. Include the location of desks, windows, doors, cabinets, and any other large items.

2. Survey your classroom for safety equipment, such as fire extinguishers and safety goggles. On your diagram, mark the location of the safety equipment. Use a different symbol for each piece of equipment. In one corner of your diagram, make a key for the symbols you use.

3. Locate the electrical outlets in the classroom. On your diagram, use symbols to mark the location of the outlets. Include the symbols in the key.

4. If your classroom has gas outlets, use a symbol to mark those on the diagram. Include the symbol in the key.

5. Mark the location of your work station. Use a colored pencil to map out the quickest and safest route for you to exit the classroom in case of an emergency.

6. On your diagram, mark any other information that you think is important.

Questions and Conclusions

1. Suppose the route you marked for a safe exit from the classroom becomes blocked. What other route might you take?

   _____

   _____

2. What safety equipment is available in your classroom?

   _____

3. What additional equipment would you like included in your classroom?

   _____

Explore Further

Make a copy of the grid on the next page and use it to make a similar diagram of your home. Share your information about safety equipment and emergency exits with family members.

Safety in the Classroom, continued

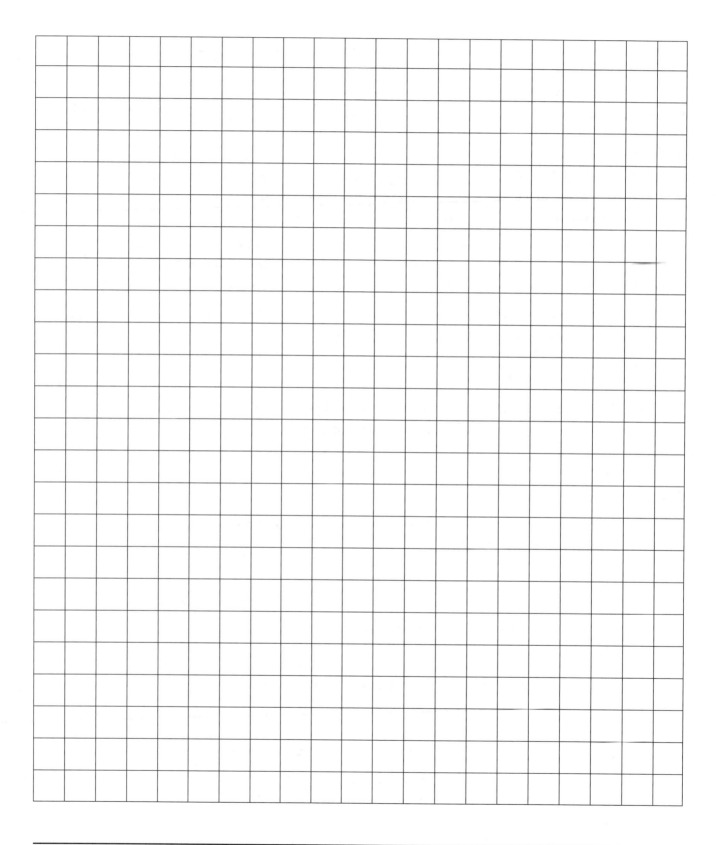

Using a Microscope

Purpose   How does a microscope help you study small things? In this
                investigation, you will use a microscope to make small objects and
                parts of objects visible.

Materials   microscope   lowercase *e* from the classified ad section of the newspaper
                microscope slide   coverslip
                eyedropper

Safety Alert Use the scissors carefully as you cut out the letter e.
                Handle the slide and coverslip carefully to avoid dropping and breaking them.
                If you break a slide or coverslip, do not pick up the broken glass with your
                hands. Use a whisk broom and dustpan to clean up the broken glass.
                Wipe up spilled water immediately.

Procedure

1. Your teacher will show you how to carry your microscope and place it on a solid
surface. The object you look at under a microscope is called a specimen. Your
microscope will have a light source to direct the light onto the specimen.

2. The first specimen that you will look at is a single cutout letter *e*.

3. Place the letter *e* on the slide in a drop of water. Make sure that the letter is
right side up so that when you look at it on the slide, you are able to read it.

4. Place the slide on the stage of the microscope.

5. Center it so that when the light passes through it, the light
will focus on the objective marked 10X. This is low power.

6. When you look through the eyepiece at
the top of the microscope and move the slide
just a little bit, you will see that the letter e just
about fills the whole field of view,
or what you can see.

7. Next, in the circle at the right, draw
exactly what you see.

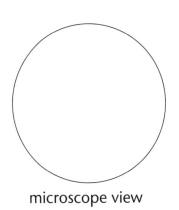

e

naked eye
view

microscope view
(10X) × (10X) = 100X

Using a Microscope, continued

Questions and Conclusions

1. Why do you think the letter e was selected?

_____

_____

2. What did the letter look like under the microscope?

_____

_____

Explore Further

Cut out another letter such as c or g. Make a drawing that shows how you think the letter will look under the microscope. Then check your prediction by using the microscope to look at the letter.

Using a High-Power Microscope

Purpose   How does a high-power microscope differ from a regular microscope? In this investigation, you will use the microscope to view greater detail and to understand the meaning of 400X magnification.

Materials   compound microscope   coverslip, microscope
microscope slide   lowercase *e* from the classified ad section of the eyedropper
newspaper

Safety Alert   Use the scissors carefully as you cut out the letter e.

Handle the slide and coverslip carefully to avoid dropping and breaking them.

If you break a slide or coverslip, do not pick up the broken glass with your hands. Use a whisk broom and dustpan to clean up the broken glass.

Wipe up spilled water immediately.

Procedure

1. To view an object under the high-power objective, you must first have the specimen in perfect focus under low power. This is true for every object that you look at.

2. Once the specimen (letter e) is in focus and you have made a record of what you see, you are ready to switch to high power.

3. Use your thumb and forefinger to turn the revolving nosepiece until it clicks into position. Now, use the fine adjustment to bring some part of the letter e into focus. You should see only a part of the letter because the magnification is so great.

4. In the circle below, draw exactly what you see in the field of view. Then, think about where the rest of the letter *e* might be outside the field of view.

e

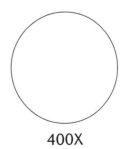

naked eye   100X   400X

Using a High-Power Microscope, continued

Questions and Conclusions

1. What do you see when you make your observations under 400X?

_____

_____

2. A What happens when you move the slide to the left?

_____

B What happens when you move the slide to the right?

_____

3. A Move the slide away from you. What happens to the image?

_____

B Now move it toward you. What happens?

_____

Explore Further

If you move the fine adjustment knob, you can see points at different levels inside the letter e print. What can you say about the "depth" of the letter e?

1 Comparing Cells

Use with Investigation 1, pages 10-11.

Purpose   Are there any differences among different types of cells?
In this investigation, you will observe differences and
similarities among different types of cells.

Draw the cells you observe under the microscope

Questions and Conclusions

1. What were some similarities between the plant cells and the animal cells?
What were some differences?

_____

_____

_____

2. How did the plant cells and animal cells differ from the bacterial cells?

_____

_____

_____

Explore Further

Observe prepared slides of other cells or find out how to make your
own slides of cells. Then draw the individual cells you observe and
compare them.

2 Living or Nonliving?

Use with Investigation 2, pages 29-30.

Purpose What are the differences between living and nonliving things? In this investigation, you will practice classifying things as living or nonliving.

Objects and properties listed will vary.

Picture 1	
Object	Properties

Picture 4	
Object	Properties

Picture 2	
Object	Properties

Picture 5	
Object	Properties

Picture 3	
Object	Properties

Living or Nonliving? continued

Picture	Nonliving Things	Living Things
1		
2		
3		
4		
5		

Questions and Conclusions

1. List three living things that you observed in the pictures.

_____

_____

2. List three nonliving things that you observed in the pictures.

_____

3. Did any of the nonliving things carry out a basic life activity, such as growing? Give an example. Then explain why you decided that it was a nonliving thing.

_____

_____

_____

4. Did any of the pictures show only living things? List the living things in the picture.

_____

_____

Explore Further

Copy the table above again. Look around you and classify everything you see as living or nonliving. Write the name of each object in the correct column.

Table 6	
Object	Properties

A Look at the Euglena

Purpose   What are the characteristics of a euglena? In this investigation, you
will observe and draw a euglena.

Protists are found in the soil, in freshwater, and in the ocean. One of the
most common and interesting of the protists is the euglena. The euglena has
a whip-like tail and a body that shortens and lengthens as it moves from
place to place. It has green parts like those found in plants, and it makes its
own food. Relatives of the euglena take food from decaying matter, and
some capture and eat bacteria.

Materials   100-power microscope   coverslips
depression slides   euglena culture
medicine dropper

Safety Alert   • Have paper towels available so that any spills can be wiped up immediately.
• Remind students that they should not pick up broken slides or coverslips.
• Make sure students wash their hands after the investigation.

Procedure

1. Use a medicine dropper to put a drop of material from the euglena culture
   onto your microscope slide. Cover with a coverslip.

2. Observe the euglena.

3. See if you can find the flagellum (tail), vacuole, eyespot, gullet, nucleus, and chloroplast.
   Refer to the picture on the next page to identify them.

4. If your euglena moves so quickly that it constantly moves out of the field of vision,
   you might want to slow it down with some gelatin, methyl cellulose, or lens paper fibers.

Directions Draw a picture of your euglena. Label the body parts that you see.

A Look at the Euglena, continued

Questions and Conclusions

1. How does your euglena look different from the one shown below?

_____

_____

2. The euglena has both plant-like and animal-like characteristics.

   A How is the euglena like a plant?

_____

   B How is the euglena like an animal?

_____

3. In what kingdom do euglena belong? Why?

_____

4. What is the purpose of the flagellum?

_____

_____

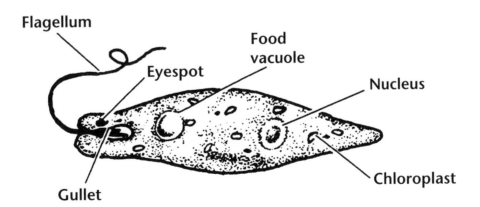

Flagellum

Eyespot

Food vacuole

Nucleus

Chloroplast

Gullet

Explore Further

A euglena can eat other protists and monerans to get energy. It can also make its own food. Observe your euglena in darkness and in light. How does it respond to the light?

Why?

Comparing Single-Celled Organisms

Purpose   How are protists such as amoebas and monerans such as a single-celled bacteria alike. In this investigation, you will compare and contrast two single-celled organisms.

Materials   high-power microscope   prepared slide of single-celled protist
prepared slide of single-celled moneran

Safety Alert • Remind students that they should not pick up broken slides or coverslips.

• Make sure students wash their hands after the investigation.

Procedure

1. Observe the protist.

2. Sketch the protist and any organelles you see. Write the name of the protist.

3. Observe the moneran.

4. Sketch the moneran and any organelles you see. Write the name of the moneran.

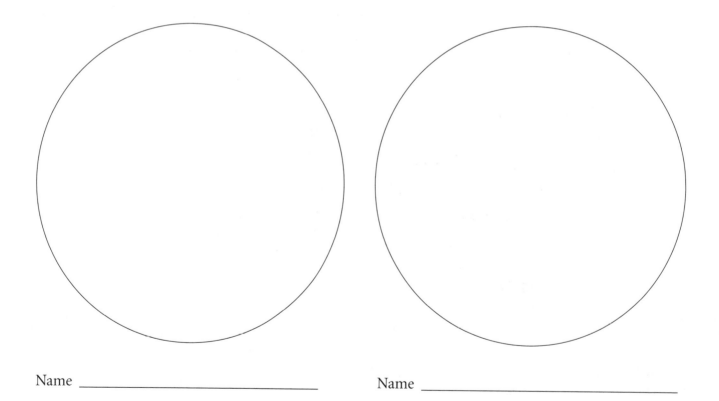

Name _____   Name _____

Comparing Single-Celled Organisms, continued

Questions and Conclusions

1. How was the protist shaped?

_____

2. What was the shape of the moneran?

_____

3. Did you see any organelles in the protist and moneran?

_____

_____

4. How are the protist and moneran alike and different?

_____

_____

_____

Explore Further

Study two other single-celled protists and monerans under the microscope.
Would you answer Question and Conclusion item 4 in the same way for
these two organisms? Explain.

Classifying Sports Equipment

Purpose   In what different ways can you classify one group of objects? In this investigation, you will make classification systems for different kinds of sports equipment.

Materials   baseball   golf ball   racquet ball   tennis ball
    baseball bat   golf club   racquetball racquet   tennis racquet

Procedure

1. Think about ways these pieces of sports equipment could be classified.

2. Invent a classification system to divide the equipment into four groups. In the boxes below, name each group and list the equipment it includes.

Name	Name
Equipment	Equipment

Name	Name
Equipment	Equipment

3. Invent a new classification system to divide the equipment into two groups. In the boxes at the right, name each group and list the equipment it includes.

Name
Equipment

Name
Equipment

Classifying Sports Equipment, continued

Questions and Conclusions

1. How many pieces of equipment were in each group in your first classification system?

_____

2. How many pieces of equipment were in each group in your second classification system?

_____

3. Suppose you were buying equipment for a baseball team at a store. Which classification system would make it easier to find what you needed at the store? Explain why.

_____

Explore Further

Classify these pieces of baseball equipment into different groups. How many groups did you use? How could you break one of these groups into two subgroups?

baseball, bases, bat, batting helmet, caps, catcher's mask, catcher's mitt, chest protector, shoes, uniform

3 Classifying Objects

Use with Investigation 3, pages 49-50.

Purpose How are objects classified? In this investigation,
you will make a classification system for objects found in
your classroom.

Questions and Conclusions

1. What were the names of the groups your team came up with?

_____

_____

2. How many levels did your classification system have?

_____

_____

3. How did your classification system differ from the systems of other student teams?

_____

_____

_____

4. How does this investigation show the value of having a single system for classifying organisms?

_____

_____

_____

_____

Explore Further

Work with two other teams to combine your classification systems into one
system. Write the combined system on a sheet of paper. Describe how your
classification system changed.

Studying Vertebrates and Invertebrates

Purpose What structures help identify animals as vertebrates or invertebrates? In this investigation, you will compare the structures of a vertebrate and an invertebrate.

Materials safety goggles small, cooked bony fish paper towels
latex gloves tweezers cooked clam

Procedure

1. Place the fish on some paper towels.

2. With the paring knife, carefully slit the skin in the middle of the back of the fish. Use the tweezers to peel the skin away.

3. Expose the backbone and skull of the fish by removing the flesh that covers them.

4. Draw a picture of the fish's backbone and skull in the box below.

5. Move the fish to one side of your work area.

6. Place a clam on some paper towels. Carefully pry apart the two pieces of the clamshell. Draw a picture of the inside of the clam in the box below.

Studying Vertebrates and Invertebrates, continued

7. Using your tweezers, search the inside of the clam for a backbone or other hard structure.

Questions and Conclusions

1. Is the fish a vertebrate or an invertebrate?

_____

2. Describe the backbone of the fish.

_____

_____

3. What other parts of the fish's skeleton were attached to the backbone?

_____

_____

4. Describe the skull of the fish.

_____

_____

5. Did you find a backbone or other hard structure inside the clam?

_____

6. Is the clam a vertebrate or an invertebrate?

_____

7. What is the main difference between vertebrates and invertebrates?

_____

_____

8. The skeleton of a fish helps protect some of the soft parts of a fish's body.
What structure helps protect the soft parts of a clam's body?

_____

Explore Further

Feel your own backbone with your hands. Observe the backbone of a
friend as he or she touches toes, twists, and bends sideways.
How is your backbone like that of the fish?

4 Identifying Angiosperms and Gymnosperms

Use with Investigation 4, pages 75-76.

Purpose What can leaves tell you about plants? In this investigation, you will identify angiosperms and gymnosperms by looking at their leaves.

Leaf	Shape	Angiosperm or gymnosperm	Vein pattern	Monocot or dicot
Leaf 1				
Leaf 2				
Leaf 3				
Leaf 4				
Leaf 5				

Questions and Conclusions

1. How do the leaves of gymnosperms differ from the leaves of angiosperms?

_____

2. Compare the shapes of the leaves of monocots to the leaves of dicots.

_____

_____

3. After you were done, did any empty boxes remain on your table? Why?

_____

_____

4. What are some other ways that the leaves you examined are different?

_____

_____

Explore Further

Make a leaf collection. Collect leaves from houseplants and plants in your yard, or ask permission to collect leaves in a garden store or a park. Dry the leaves between layers of newspaper stacked under heavy books. Mount each leaf on a sheet of paper. Identify each leaf as you did in the investigation.

Tree Study

Purpose   What attributes should you examine when you observe a tree?
In this investigation, you will observe and describe a tree.

Materials   meat thermometer   hand lens   air thermometer
paper   tape measure   crayons
trowel (small garden shovel)   latex gloves

Safety Alert • Check for allergies before students begin this investigation.
Have students wear gloves when examining the soil. Some students
may be allergic to latex.

• Use unbreakable thermometers.

• Show students the proper way to use the trowel.

• Caution students not to touch their faces with their gloves.

• Make sure students wash their hands carefully after the investigation.

Procedure

1. Take a walk around the school with your class. Choose a tree. Observe it carefully.
Find out more about it. Come back and observe it in another season.

2. Use a meat thermometer to measure the temperature of the soil around your tree. Hang an air
thermometer on a branch to measure the air temperature. Leave each thermometer in place for five
minutes before reading it.

Soil temperature _____ Air temperature _____

3. Try to encircle the tree with your arms. Use the tape measure to measure the circumference of the
tree trunk. The circumference is the distance around the tree trunk.
Circumference _____

4. Identify the color of the tree's bark. _____

5. Feel the bark. Describe its texture: smooth, rough, and so on. Make a bark rubbing by putting a sheet
of paper on the bark and rubbing the paper with the side of a crayon.

_____

6. If your tree has leaves, describe them. Note their texture and color. Make a leaf rubbing as you did a
bark rubbing, or sketch and color the leaf.

_____

7. Look at the top of the tree. This part is the tree's crown. Place a long tape measure on the ground
and estimate the diameter of the crown. The diameter is the distance from one side to the other.
Record the diameter of your tree's crown.

_____

Tree Study, continued

8. Look at the shape of your tree from the bottom to the top and from one side to another. Describe the shape of your tree.

_____

9. List any of the following items that are on your tree: bud, seed, fruit, moss, fungi, or other items.

_____

_____

10. Use a trowel and hand lens to investigate the material at the base of the tree. Do you see moss, grass, weeds, needles, or leaves? List all the things that you find.

_____

_____

11. Do you see signs of insects, birds, squirrels, or other small animals? Make another list of animals or signs of animals using your tree.

_____

_____

Questions and Conclusions

1. In the tree an angiosperm or a gymnosperm?

_____

2. What is the species of the tree? _____

3. Is the tree a young or a mature tree? Explain your answer.

_____

Explore Further

What do you think your tree will look like in another season? Describe how it may differ from the way it looks now.

Dicot and Monocot Seeds

Purpose   How are monocot and dicot seeds alike and different? In this investigation, you will observe and compare and contrast dicot and monocot seeds.

Materials   presoaked lima beans and corn seeds
              hand lens
              knife
              paper towels
              stereomicroscope (if available)

Safety Alert • Cut open enough corn seeds for each student.

              • Make sure students wash their hands carefully after the investigation.

Procedure

1. Get a lima bean seed and a corn seed from your teacher.
Place them on a paper towel.

2. Carefully open the lima bean seed with your fingers so that the two halves are separated. Find the embryo leaves.

3. Observe the seed with the hand lens. Draw a picture of what you see. Label your illustration.

4. Obtain a corn seed that is cut open from your teacher. Find the embryo leaf.

5. Observe the seed with the hand lens. Draw a picture of what you see. Label your illustration.

6. If you have one, observe the seeds with a stereomicroscope.

Dicot and Monocot Seeds, continued

Questions and Conclusions

1. Which seed is from a plant that is a monocot? How can you tell?

_____

2. Which seed is from a plant that is a dicot? How can you tell?

_____

3. What is each half of the split bean called?

_____

Explore Further

Why do you think the cotyledons in lima beans are nutritious food for humans?

_____

_____

Observing Paramecia

Purpose   How does the contractile vacuole in a paramecia function? In this investigation, you will study the function of contractile vacuoles in paramecia.

Materials
(per group):

safety goggles	latex gloves	
medicine droppers	methyl cellulose	
microscope slide	paramecium culture	
coverslip	compound light microscope	
stopwatch	distilled water	
paper towel	salt water	

	Number of contractions per minute		
	Culture medium	Distilled water	Salt water
Paramecium #1			
Paramecium #2			
Paramecium #3			
Average			

Procedure

1. Wear your gloves throughout the investigation. Do not touch your face with your gloves or eyedroppers. Wipe up any spills immediately.

2. Have your teacher use an eyedropper to place a paramecia culture on your slide. Cover the specimen with the coverslip.

3. Place the slide under the microscope and focus so that you can view the contractile vacuole of a paramecium.

4. As a group member uses the stopwatch to time one minute, count the number of times the vacuole contracts. You may have to reposition the slide to keep the contractile vacuole in view while counting the contractions. Record your count on the chart. Repeat the procedure as you watch two other paramecia.

5. Fill medicine droppers with distilled water and salt water solutions from your teacher. The teacher has prepared a 100-mL water solution with 10 g of methyl cellulose with distilled water and a 100-mL salt solution with 5 g of sodium chloride and distilled water.

6. Get a new specimen and add a drop of distilled water with methyl cellulose onto the slide. If the slide becomes dry, add more distilled water to the slide. Repeat step 4.

Observing Paramecia, continued

7. Get a new specimen and add a drop of salt water to the slide. If the slide becomes dry, add more salt water to the slide. Repeat step 4.

8. Find the average number of contractions in each solution.

Questions and Conclusions

1. How did distilled water affect the activity of the contractile vacuoles?

_____

_____

2. How did salt water affect the activity of the contractile vacuoles?

_____

_____

3. In distilled water, the concentration of water is higher than it is in culture medium. In salt water, the concentration of water is lower than it is in culture medium. Using this information, explain the effects of distilled water and salt water.

_____

_____

_____

_____

Explore Further

Repeat the investigation using a different solution such as a sugar or weak-soap solution. Compare your results to the original findings. What can you conclude about the effects of the new solution on the paramecia?

Studying Yeast

Purpose   Does yeast grow better at high or low temperatures? In this investigation, you will study the effect of temperature on the growth of yeast.

Materials: safety glasses   thermometer
3 test tubes   microscope slide
package of dry yeast   iodine solution
3 medicine droppers   coverslip
maple or corn syrup   compound light microscope
refrigerator

Procedure

1. Use the table below to record your data.

Storage site	Temperature (°C)	Number of cells
Refrigerator		
Cabinet		
Warm place		

2. Put on the safety glasses.

3. Fill each test tube about half-full of water. Divide the yeast from the package into thirds. Place one-third of the yeast in each test tube.

4. Use a medicine dropper to add 10 drops of syrup to each test tube. Gently swirl the tubes to mix the contents.

5. Store one tube in a refrigerator, one in a cabinet at room temperature, and one in a warm place, such as near a heater vent.

6. The next day, use the thermometer to measure the temperature of the yeast culture in each tube. Record the temperatures in your data table.

7. Gently swirl the refrigerated tube to mix the contents. With a new medicine dropper, transfer a drop of yeast culture from that tube to a microscope slide.

8. Use a clean medicine dropper to add a drop of iodine solution to the slide. Cover with a coverslip.

9. Place the slide on the microscope stage. Count the yeast cells that appear in one field of view under high power. Record the number in your data table.

10. Repeat steps 7–9 for the other two tubes. Wash the medicine dropper with water before letting it touch each yeast culture. Record the numbers in your data table.

Studying Yeast, continued

Questions and Conclusions

1. How do yeast cells reproduce?

_____

2. Which of the three yeast cultures had the most cells?

_____

3. Which of the three yeast cultures had the fewest cells?

_____

4. What do your results indicate about the effect of temperature on the growth of yeast?

_____

_____

_____

Explore Further

Have students test the effect of other variables on the growth of yeast. They might consider varying the amount of sugar in the culture or varying light, for example. Have students draw conclusions based on their observations.

5 Growing Bread Mold

Use with Investigation 5, pages 105-106.

Purpose   In this investigation, you will study the effects of light and moisture on the growth of mold. Does mold grow faster on wet or dry bread? How does light or its absence affect mold growth?

Materials   slice of dried white bread   stereomicroscope
masking tape   4 petri dishes with lids
table knife

Procedure

1. Use the table below to record your data.

	Appearance of bread		
Storage condition	**After 1 day**	**After 2 days**	**After 3 days**
Dry bread in light			
Moist bread in light			
Dry bread in dark			
Moist bread in dark			

Questions and Conclusions

1. Which dish had the most mold growth after three days?

_____

2. Which dish had the least mold growth after three days?

_____

3. What do your results indicate about the effects of light and moisture on the growth of mold?

_____

_____

_____

Growing Bread Mold, continued

4. Why was it important to let the dishes sit uncovered for 30 minutes before sealing them?

_____

_____

_____

5. Why was it important that the lighted place should not be warmer or colder than the dark place?

_____

_____

_____

Explore Further

Plan an investigation to study the effects of another environmental
condition on mold growth. Write a purpose, materials list, and procedure.
Then perform your investigation.

Observing the Action of a Digestive Enzyme

Purpose   What happens to food as it is chewed? In this investigation, you will observe how an enzyme in saliva can break down starch.

Materials: laboratory gloves   mortar and pestle
soda cracker   4 small paper cups
2 medicine droppers   iodine solution
cooked egg white

Procedure

1. Use the table below to record your data.

Condition	Color after Adding Iodine Solution
Ground cracker	
Chewed cracker	
Ground cracker + water	
Ground egg white	
Chewed egg white	
Ground egg white + water	

2. Put on your gloves. Use a mortar and pestle to grind a soda cracker into a fine powder. Put the ground cracker in a paper cup.

3. Chew a second cracker thoroughly so that it is completely mixed with saliva.

4. **Safety Alert: Do not touch the chewed cracker with your hands.** Spit the chewed cracker into a second paper cup. Be sure that every part of the cracker is mixed with saliva. If necessary, spit into the cup to get more saliva on the cracker.

5. Repeat step 2 using a third paper cup. Then use a medicine dropper to add several drops of water to this cup. Add enough water to make the ground cracker as moist as the chewed cracker in step 4.

6. Use another medicine dropper to add a drop of iodine solution to each cup. Iodine solution reacts with the carbohydrate known as starch. This reaction causes a blue-black color to appear.

Observing the Action of a Digestive Enzyme, continued

7. Observe the color of the mixture where the iodine solution was placed in each cup. Record your observations in your data table.

8. Repeat steps 2–7 using pieces of cooked egg white. Do not chew the cooked egg if you are allergic. Observe another student's chewed egg.

Questions and Conclusions

1. Do your results indicate that soda crackers contain starch? Explain why or why not.

_____

_____

2. Do your results indicate that egg whites contain starch? Explain why or why not.

_____

_____

3. Saliva contains a digestive enzyme that breaks down starch. Use this information to explain why one of the cups did not have a blue-black color.

_____

_____

4. Why was it important to test a cup containing a ground cracker and water?

_____

_____

Explore Further

Put a slice of cucumber and a slice of banana on a plate. Add several drops of iodine solution to each. What results do you observe? What can you conclude about these foods? What happens if you put saliva on the banana slice?

6 Studying Feeding in Hydras

Use with Investigation 6, pages 121-122.

Purpose   What stimuli trigger feeding responses in a simple animal? In this investigation, you will observe feeding behavior in a hydra and record its responses to different stimuli.

Stimulus	What happened?
Water fleas	
Moving filter paper	
Touched by filter paper	
Beef broth on filter paper	

Questions and Conclusions

1. How were the hydra's responses to water vibrations, touch, and chemicals different?

_____

_____

_____

_____

2. Which of these stimuli probably triggers feeding responses when a hydra consumes a water flea?

_____

Explore Further

Look for dead water fleas on the bottom of their culture dish. Find out if a dead water flea will trigger a feeding response in a hydra. Do the results of this test support your answer to question 2?

Measuring Carbon Dioxide Production

Purpose   When do your lungs release the most carbon dioxide?
In this investigation, you will measure the amount of carbon dioxide
in exhaled air.

Materials: safety glasses   plastic wrap
100 mL graduated cylinder   grease pencil
4 flasks (250 mL)   drinking straw
0.1% methyl red   0.04% sodium hydroxide
2 medicine droppers

Procedure

1. Use the table below to record your data.

Condition	Drops of Sodium Hydroxide Added
After normal breathing	
After holding breath	
After exercising	

2. Put on the safety glasses.

3. Use the graduated cylinder to measure 100 mL of water. Pour the water into one of the flasks.

4. Use a medicine dropper to add 10 drops of methyl red to the flask. Swirl the flask gently.

5. Cover the flask with plastic wrap. Use the grease pencil to write "Control" on the side of the flask.

6. Repeat steps 3 and 4 with a second flask.

7. Exhale gently through a straw into the second flask for exactly 2 minutes. The carbon dioxide in your breath will enter the solution. That will make the solution change color. **Safety Alert: Do not inhale the solution.**

8. Use another medicine dropper to add sodium hydroxide to the second flask. Count the drops as you add them. Swirl the flask after each drop.

9. Sodium hydroxide will make the solution turn pink again. Stop adding drops when the color of the solution in the flask matches the color of the "Control" flask. In your data table, record the number of drops you added.

Measuring Carbon Dioxide Production, continued

10. Repeat steps 3 and 4 with the third and fourth flasks. Cover both flasks with plastic wrap.

11. Hold your breath for 30 seconds. Immediately uncover the third flask and repeat steps 7–9.

12. Exercise by doing jumping jacks for 2 minutes. Immediately uncover the fourth flask and repeat steps 7–9.

Questions and Conclusions

1. Which flask needed the most drops of sodium hydroxide to match the color of the "Control" flask?

_____

2. Which flask needed the fewest drops of sodium hydroxide to match the color of the "Control" flask?

_____

3. The more carbon dioxide a solution contains, the more sodium hydroxide you must add to make it turn pink. Use this information to rank the three flasks according to the amount of carbon dioxide they contained.

_____

_____

4. Why did holding your breath cause you to exhale more carbon dioxide than normal?

_____

_____

5. Why did exercising cause you to exhale more carbon dioxide than normal?

_____

Explore Further

How could you measure the amount of carbon dioxide in the air in your classroom? Do you predict it would have a higher or a lower amount of carbon dioxide than is normally present in outside air? Why?

Vascular Tissue

Purpose   How do fluids behave in the stem of a plant? In this investigation, you will compare the action of fluids in a celery plant to those in a white carnation.

Materials   1 celery stalk with leaves
1 white carnation
2 regular water glasses
red or blue food coloring
knife or scalpel

Procedure

1. Put two to three inches of tap water into each glass. Add a few drops of food coloring to each water glass.

2. **Safety Alert: Use caution when handling knives or other cutting tools.** Cut off the bottom end of the stem of the carnation or have your teacher cut off the end. Look at it closely. Place the carnation in one of the glasses.

3. Cut off the bottom end of the celery stalk or have your teacher cut it. Look at it closely. Place the celery stalk in the other water glass.

Questions and Conclusions

1. Describe the appearance of the plants before you placed them in the colored solution.

_____

_____

2. What was the purpose of making a fresh cut on the stem?

_____

_____

3. Describe the appearance of the plants now. How would you explain what happened?

_____

_____

Vascular Tissue, continued

4. Remove each of the stems from the colored solution the next day. Make a new cut. Look closely at the freshly cut piece of the stem. Describe what you see.

_____

_____

5. Draw a diagram showing what took place.

Explore Further

Cut thin sections of the celery stalk and examine them under a microscope.
Find the vascular tissue. What does it look like?

Sprouting Seeds

Purpose   What happens when seeds germinate? In this investigation, you will observe sprouting seeds.

Materials   100 bean seeds
12-inch × 12-inch piece of towel
5–6 pages of newspaper
rubber bands
water

Procedure

1. Wet the cloth thoroughly but not dripping. Lay the cloth on the newspaper.

2. Arrange the seeds about one inch apart in ten rows of ten each. Roll the paper with the cloth and the seeds. Use rubber bands or string to hold the rolled paper together.

3. After the second day, unwind the roll to observe the seeds. Make a record of the number of seeds that have started to grow.

4. Sprinkle the cloth with water but do not soak it.

5. Observe the seeds for the next ten days. Fill in the bar graph with the information you record over a ten-day period.

Sprouting Seeds

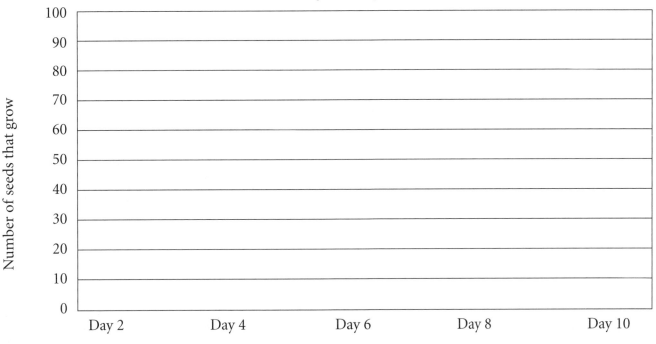

Sprouting Seeds, continued

Questions and Conclusions

1. Write a paragraph to describe what you observed in this investigation during the ten-day period.

_____

_____

_____

2. Look at your bar graph. Do you see a pattern in the data? Write a few sentences to summarize the data you collected.

_____

_____

_____

Explore Further

Repeat the investigation. This time, use two different groups of seeds. Provide more heat and light for one group than for the other. What differences do you see between the two groups?

7 Growing an African Violet From a Leaf

Use with Investigation 7, pages 161–162.

Purpose Can one leaf grow into a whole new plant? In this investigation, you will grow a plant, using asexual reproduction.

Questions and Conclusions

1. What was the first change that you observed in the leaf? Describe your observation.

_____

_____

_____

2. Why do you think the plant produced this type of new growth?

_____

_____

_____

3. How does the plant change after the leaf is planted in the soil?

_____

4. What type of reproduction occurred in this investigation? Explain your answer.

_____

_____

_____

Explore Further

Many plants will grow from leaf or stem cuttings. Try to grow some other plants this way. You may want to use a book on houseplants as a reference.

How Does Exercise Change Heart Rate?

Use with Investigation 8, pages 179-180

Purpose Does your heart beat faster when you're very active?
In this investigation, you will observe the changes in heart rate
during different amounts of activity.

Activity	Heart rate
Sitting (resting heart rate)	
Standing up	
After running in place	
Resting 30 seconds	
Resting 1 minute	
Resting $1\frac{1}{2}$ minutes	
Resting 2 minutes	
Resting $2\frac{1}{2}$ minutes	
Resting 3 minutes	

Questions and Conclusions

1. How does the amount of activity affect heart rate? _____

_____

2. Describe the demands on your heart when the heart rate was lowest and when it was highest.

_____

_____

3. Why does the heart rate change as the amount of activity changes? _____

_____

Explore Further

Design an investigation about heart rate. You could investigate one of the
following questions or one of your own: What activities increase heart rate
the most? How much do people's resting heart rates differ? Once you
choose a purpose for your investigation, write a procedure and do the
investigation.

Diffusion

Purpose Can substances move into and out of cells? In this investigation, you will study an example of the movement of materials in living cells.

Background Information Diffusion is a process that involves movement of molecules, including solutes and solvents, into and out of cells. Solutes and solvents move by "diffusing" across cell membranes. Diffusion is an active process in which solutes and solvents, like water, move from an area of higher concentration to an area of lower concentration. In this activity, changes in a carrot will be used to depict the movement of water by diffusion.

Materials

2 carrots	labels	60 cm thread or thin string
marking pens	salt	water
paper towels	scalpel or knife	metric ruler
stirring rod	4 beakers or jars	

Procedure

1. Cut the carrots in half so that you have two tops and two bottoms.
   Safety Alert: Use caution when using scalpels, knives, or other cutting implements.

2. You will make two setups; one with the tops of the carrots and a second with the bottoms of the carrots.

3. Tie the thread or string tightly around the carrot pieces 2 cm from the cut surface, as shown in the picture.

4. Label two of the beakers Salt Water. Fill them three-fourths full of water.
   Add 15 grams of salt to the water and stir. **Safety Alert: Wipe up spills immediately.**

5. Label the other two beakers Freshwater. Fill them three-fourths full of water.
   Safety Alert: Wipe up spills immediately.

6. Each partner will use one freshwater container and one saltwater container.
   One partner will observe the top halves of the carrots, and the other partner will observe the bottom halves.

7. Observe the carrot parts. Place one carrot part in each container.
   Describe the conditions of each carrot part as it is placed in the container.

8. Allow the carrots to remain in the containers for 24 hours. Then remove the carrots.

9. After 24 hours, describe the condition of the carrots in each container.
   Note any differences in the tightness of the threads or strings.

Diffusion, continued

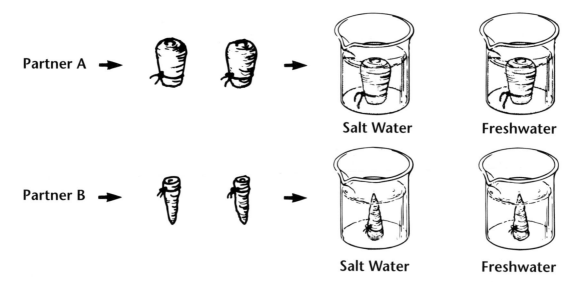

Partner A → → Salt Water   Freshwater

Partner B → → Salt Water   Freshwater

Questions and Conclusions

1 In which kind of water did the carrot cells lose water?_____

Explain._____

_____

2 In which kind of water did the carrot cells gain water?_____

Explain. _____

_____

3 What was the purpose of tying thread or string on each of the carrot pieces? _____

Explain. _____

_____

4 The movement of liquids and gases between cells depends on differences in the concentration of solutes and solvents in the cells. In which direction do liquids and gases move? Why do liquids and gases move from one concentration to another?

_____

_____

Explore Further

Give an example of substances in which diffusion might occur in the excretory and respiratory systems.

Sensory Receptors in the Skin

Purpose   How do different parts of your skin respond to various sensations? In this investigation, you will map the location of sensory receptors for pressure and temperature on the hand.

Background Information   There are several kinds of sensory receptors in your skin. Some are sensitive to touch while others are sensitive to pain. Still others sense heat or cold.

Materials   outlines of each partner's left hand   cup of ice
three thin nails   cup of hot water
paper towels

Procedure

1. Work in pairs. On each hand outline, draw ten scattered sites and label them from 1 to 10. The sites should be located in similar places on both outlines.

2. Place one nail in ice water to cool. Place another nail in the hot water to warm. **Safety Alert: Be sure that the hot water is not hot enough to cause burns. Clean up spills immediately.**

3. Gently touch the other nail to each site on your partner's hand to detect which of the sites are sensitive to pressure. Sites that feel the point more sharply are sensitive to pain. Reverse roles, having your partner test the sites on your hand.

4. Note on your outline which sites are sensitive to pressure and pain.

5. Repeat step 3 with the nail that has been in ice for several minutes. Note on the outline which sites are sensitive to cold.

6. Repeat step 3 once more with the nail that has been in hot water for several minutes. Note on the outline which sites are sensitive to heat.

Question and Conclusions

1. What makes some areas sensitive to cold and others sensitive to pain?

_____

_____

Explore Further

Repeat the investigation with eight sites on the back of the neck. Which location, hand or neck, is more sensitive to sensory stimuli? What makes you say this?

Observing Mitosis

Purpose   What do cells look like when they are in the process of dividing? In this investigation, you will examine different phases of mitosis in plant cells.

Materials   compound light microscope
prepared slide of an onion root tip

Safety Alert • Do not use sunlight as your light source. Keep electrical cords away from water and tape them to your desk to prevent a stumbling hazard.

• Handle slides carefully. If you break one, do not pick it up with your hands. Sweep the broken glass up with a whisk broom and dispose of it in a special container for glass.

Procedure

1. Review with your teacher how to properly use a compound light microscope.

2. Put your slide on the stage. Place the metal clips over the edges of the slide to hold it in place.

3. Use the low-power objective to observe the slide.

4. Carefully change to the high-power objective. Adjust the focus until you can see the cell walls and nuclei of the onion cells.

5. Refer to the figure of mitosis on page 215 of your Student Text. Look for examples of each stage of mitosis in your slide. If you cannot find an example of each stage, move your slide and begin with step 3 again.

6. In the space below, draw and label a diagram of each stage of mitosis that you find.

Questions and Conclusions

1. What shape are the onion root cells? _____

2. What color are the chromosomes stained on your slide? _____

Explore Further

Count up the number of cells you see in each phase. Classify each cell based on the phase it is in and use those numbers to predict in which phase the cells spend the most time. Create a chart and calculate percentages.

Observing Sperm and Egg Cells

Purpose   What do the sperm and egg cells of a sea star look like? In this investigation, you will examine compare sea star sperm cells with sea star egg cells.

Materials   compound light microscope   prepared slide of sea star egg cells
prepared slide of sea star sperm cells

Safety Alert   • Do not use sunlight as your light source. Keep electrical cords away from water and tape them to your desk to prevent a stumbling hazard.
• Handle slides carefully. If you break one, do not pick it up with your hands. Sweep the broken glass up with a whisk broom and dispose of it in a special container for glass.

Procedure

1. Review with your teacher how to properly use a compound light microscope.

2. Put your slide of the sea star eggs on the stage. Place the metal clips over the edges of the slide to hold it in place.

3. Use the low-power objective to observe the slide. Then carefully change to the high-power objective.

4. In the space below, draw what you see.

5. Put your slide of the sea star sperm on the stage. Place the metal clips over the edges of the slide to hold it in place.

6. Use the low-power objective to observe the slide. Then carefully change to the high-power objective.

7. In the space below, draw what you see.

Questions and Conclusions

1. Compare your drawings of the egg and sperm. How are eggs and sperm alike and different?

_____

2. What structure on the sperm allows it to move through the female reproductive system to an egg?

_____

Explore Further

What is the shape of the sea star egg? What advantages does this shape have over other shapes, such as oval or rectangular?

9 Graphing Gestation Times

Use with Investigation 9, pages 226–227.

Purpose How do gestation times of different animals compare with each other? In this investigation you will use a bar graph to compare different gestation times.

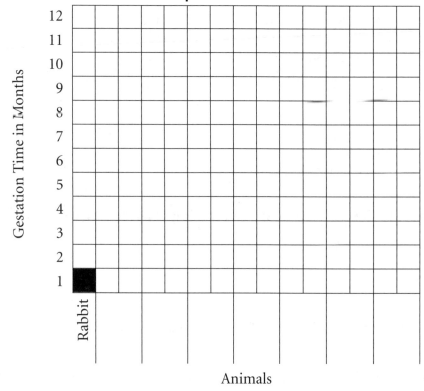

Bar Graph of Gestation Times

Animal	Months
Cat	2
Cow	10
Dog	2
Ewe	5
Goat	5
Mare	11
Rabbit	1
Sow	4

Questions and Conclusions

1. Which two gestation times are almost similar in length to the human gestation time?

_____

2. What does each line in the scale on the left side of the graph represent?

_____

Explore Further

Assume that an animal can have an egg or eggs fertilized shortly after giving birth. Calculate how many times each of the animals in your bar graph can give birth during one year (365 days). Make a new bar graph that compares the information. Explain how the length of gestation time affects the number of times an animal can give birth during one year.

Fighting Pathogens

Purpose   Does vinegar have an effect on the growth of pathogens? In this investigation, you will test the effects of an acid on pathogen growth.

Materials   2 disposable nutrient agar dishes   marker   tape
2 sterile cotton swabs   sterile distilled water   spoon
distilled white vinegar   paper towels

Safety Alert   • Do not inhale or taste the vinegar   • Clean up any spills immediately.
• Do not open your dishes after you have sealed them.
• Treat all growth as through it were a pathogen.

Procedure

1. Oil and sweat contain acids. These acids kill pathogens. In this investigation, you will test the effect of the acid vinegar on the growth of pathogens.

2. Label your agar dishes: "With vinegar" and "Without vinegar."

3. Wet one of your cotton swabs and rub it across a part of your classroom that is likely to contain pathogens, such as your desk or a doorknob. Then open the cover to one of your dishes and wipe the cotton swab across the surface of the agar. Close the dish immediately.

4. Repeat step 3 with the other agar dish.

5. Pour a spoonful of vinegar over the agar in the dish you labeled "With vinegar."

6. Seal both dishes with tape. Do not reopen the dishes.

7. Put the dishes in a warm, dark place for one week.

8. After one week, observe your dishes, but do not open them.

Questions and Conclusions

1. Compare your two dishes. Do they look alike or different? _____

2. Explain any differences you see.

_____

_____

Explore Further

How could you test how well a household cleaning product sanitizes an area? Prepare a second set of agar dishes. Swipe dish A with a cotton swab rubbed over a dirty area in the classroom. Use a cleaner to sanitize the area. Then swipe the same area and wipe the swab on the surface of dish B. Treat both dishes as you did in the investigation. What differences would you expect to see in the two dishes? Why?

10 Reading Food Labels

Use with Investigation 10, pages 254–255.

Purpose How do nutrients in different foods compare? In this investigation, you will determine the kinds and amounts of nutrients in different packaged foods.

	Food	Serving size	Carbohydrates (in grams)	Proteins (in grams)	Fats (in grams)	Vitamins and minerals (percentages)
1						
2						
3						
4						
5						
6						
8						

Questions and Conclusions

1. Compare the data in your table. Which food has the most carbohydrates?

_____

2. Which food has the most proteins?

_____

3. Which food has the most fats?

_____

4. Which food has the most vitamins and minerals?

_____

5. Compare the serving sizes of the foods. Which food has the smallest serving size? Which has the largest serving size?

_____

Reading Food Labels, continued

Explore Further

1. Look at all the data you collected. Which food do you think is best for you to eat? Give reasons for your choice.

_____

_____

2. Does the information in your table change your ideas about eating certain foods? Why or why not?

_____

_____

Comparing Calories Used in Exercise

Purpose   How much energy do you use by doing different activities? In this investigation, you will compare the number of calories needed for specific activities.

Materials   pencil or colored pencils

Procedure

1. Given below is the number of calories per hour that certain activities use.

2. Use the information in the table to complete the bar graph. Put the activity using the least amount of calories in the first bar. Continue so that the activity using the most calories is in the last bar. The first activity has been graphed for you.

Activity	Calories used
Bicycling, moderate	450
Jogging, 5 mph	500
Swimming, active	500
Sitting	85
Walking, 3 mph	240
Standing	100

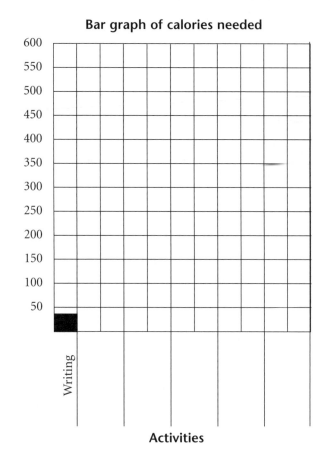

Bar graph of calories needed

Activities

Questions and Conclusions

1. Which two activities use the most calories?

_____

2. What is similar about sitting and standing?

_____

3. Which activity surprised you the most? Explain.

_____

Explore Further

Power walking uses up 600+ calories an hour. Skipping rope uses up 700+ calories per hour. Which of these two would you prefer for a weight-loss program? Why?

Modeling Mendel's Experiments

Purpose   What did Mendel prove in his experiments with pea plants? In this investigation, you will model the genetic crosses that resulted in Mendel's F_2 generation.

Materials   2 purple beads
             2 white beads

Procedure:

1. You will need one partner for this investigation. Each partner should have one purple bead and one white bead. The purple bead represents a gene for purple flowers in pea plants. The white bead represents a gene for white flowers in pea plants.

2. Close your hand around the two beads and place your hand behind your back. Shake the beads. Your partner does the same thing.

3. Without looking at the beads, each partner chooses a bead. Look at the beads.

4. Choose a letter to represent flower color in pea plants. Then record the results of the cross in the table below.

Cross	My gene	My partner's gene	Genotype	Phenotype
1				
2				
3				
4				
5				
6				
7				
8				
9				
10				

5. Repeat steps 1–4 nine more times, filling in the chart as you go.

6. After you have recorded the results of ten crosses, write the genotype and phenotype of each offspring in the table.

Modeling Mendel's Experiments, continued

Questions and Conclusions

1. What is the genotype and phenotype of the parents in the crosses you did?

_____

_____

2. Draw a Punnett square to represent this cross.
What are the chances that the offspring will have purple flowers?

_____

Explore Further

Do the results of your investigation confirm the information on the
Punnett square? Explain why or why not.

Modeling Sex Determination

Purpose   How do the sex chromosomes of the parents determine the sex of a baby? In this investigation, you will model the number of boys and girls born into ten families.

Materials   a nickel
a dime
masking tape
marking pen

from father

from mother

Procedure

1. In this investigation, you will determine the sex of the babies in ten families. Use a nickel to stand for the male chromosome pair. Put a piece of masking tape on each side of the coin. Write X on one side. Write Y on the other side.

2. Let a dime stand for the female chromosomes. Put a piece of masking tape on each side of the coin. Write an X on each side. Females have only X chromosomes.

3. Each family will have four children, so you will flip the coins four times for each family, or 40 times altogether. Start with Family #1. Flip the two coins.

4. Observe the results of your coin flip. If they come out XX, it's a girl. Put a mark in the box for Girls of Family #1. If the coins come out XY, put a mark in the box for Boys for Family #1.

5. Flip the coins three more times to determine the sex of the other three children in Family #1.

6. Repeat steps 3–5 nine times, for a total of 40 coin flips. Record the sex of each child in the data chart.

7. Add the total number of girls and boys produced and write the number in the data chart.

Family #	1	2	3	4	5	6	7	8	9	10	Total
Number of girls											
Number of boys											

Modeling Sex Determination, continued

Questions and Conclusions

1. How many girls did you get out of the 40 flips?

_____

2. How many boys did you get out of the 40 flips?

_____

3. What percentage of the babies are girls?

_____

4. What percentage of the babies are boys?

_____

5. Compare your results with other class members. How do your results compare with theirs?

_____

6. What does this tell you about the possibility of having girls or boys in a family?

_____

Explore Further

Suppose a family has five children, all boys. Are the chances of a sixth child being a girl greater than those of the families in your chart?
Tell why or why not.

11 Tracing a Genetic Disease

Use with Investigation 11, pages 286–287

Purpose How can a genetic disease in a family be traced from generation to generation? In this investigation, you will trace a genetic disease by using a family history and a diagram called a pedigree

Pedigree of John's Family

Questions and Conclusions

1. Based on the pedigree on page 286, is albinism caused by a recessive gene or by a dominant gene? Explain.

_____

_____

2. Look at the pedigree you drew in step 4. How many generations does your pedigree show?

_____

3. Does the pedigree you drew show a genetic disease that is caused by a recessive gene or by a dominant gene? How do you know?

_____

_____

4. From which parent did John inherit Huntington's disease?

_____

5. Do you think that Nancy is a carrier for Huntington's disease? Why or why not?

_____

Tracing a Genetic Disease, continued

Explore Further

1. Based on your answer to question 3 above, give John's genotype for Huntington's disease. Explain your answer.

2. Assume that John marries a woman who does not have Huntington's disease. Draw a Punnett square to show the chance that a child of theirs will inherit the disease.

Life in a Pond

Purpose   Is there life in a drop of pond water? In this investigation, you will observe the microorganisms of a pond ecosystem.

Background Information   Microorganisms are common in pond and swamp communities. Although many of these animals have only one cell, they carry out the basic life activities.

Materials

hand lens	jar of pond water	paper towels	depression slides
microscope	protective gloves	coverslips	medicine dropper

Safety Alert   Wear protective gloves when handling the pond water.
Do not pick up broken slides with your hands.
Clean up spilled water immediately.

Procedure

1. With your medicine dropper, put a small drop of pond water on your slide. Cover the water drop with a coverslip. Observe the water drop with a hand lens.

2. Look at the slide with a microscope. If you do not see any microorganisms, move the slide until you do. If you still cannot locate an organism, make a new slide.

Questions and Conclusions

1. Describe some of the microorganisms and how they are behaving.

   _____

2. In the circles, sketch two different microorganisms that you see in the field of the microscope.

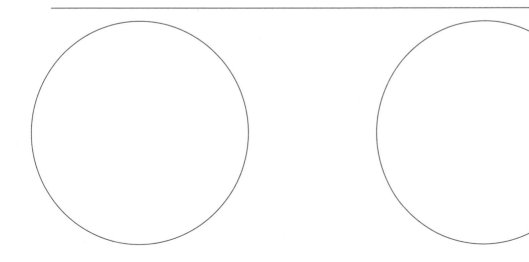

Life in a Pond, continued

Explore Further

An ecosystem has living things that interact with each other and with the nonliving parts of the environment. Think about the pond ecosystem you visited. Name the living and nonliving parts of this pond ecosystem. How do the microorganisms interact with them?

12 Testing the pH of Rain

Use with Investigation 12 on pages 304–305.

Purpose How can you tell if rain is acid rain? In this investigation, you will find out if the rain in your area is acid rain.

Date	pH of sample 1	pH of sample 2	pH of sample 3	Average pH of samples
pH of distilled water:				

Questions and Conclusions

1. How does the pH of distilled water compare with the pH of the rainwater you tested?

_____

_____

2. When water has a pH lower than 7, it is acidic. Normal rain is always slightly acidic and has a pH between 4.9 and 6.5. When rain has a pH of less than 4.9, it is called acid rain. Were any of the samples of rainwater you tested acid rain?

_____

_____

3. Was the pH of the rainwater the same every day you collected samples? What are some reasons the pH of rainwater could vary from day to day?

_____

_____

Explore Further

Test the pH of the water in a local pond, lake, or stream. Safety Alert: Wear protective gloves when you collect the water samples. Is the pH of the body of water the same as that of rainwater? Why might the values be different?

A Food Chain

Purpose   How does energy flow through the organisms in a freshwater
                     environment? In this investigation, you will study the relationships
                     that exist in a freshwater ecosystem?

Background   There is an interdependence among animals, plants, and protists that live in
Information   localized communities. Usually the dominant organism gives the system its name.

                     A pickerel is a large freshwater fish that may depend upon smaller fish, such as the
                     perch, for its main food source. The perch depends on yet smaller fish.

Materials   pencil   encyclopedia   paper   dictionary

Procedure

   1. Arrange the following organisms in an order that shows how energy might flow from
          organism to organism.

Start here

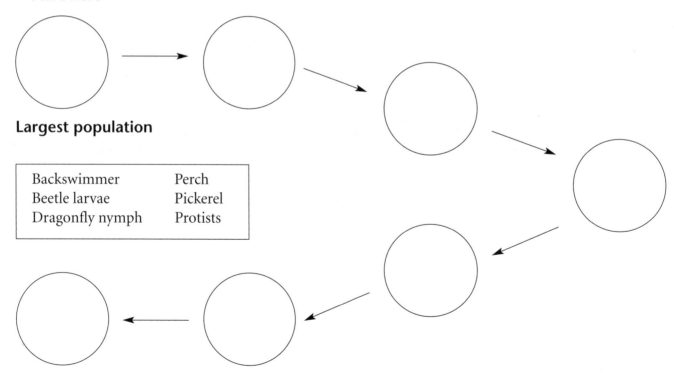

Largest population

Backswimmer	Perch
Beetle larvae	Pickerel
Dragonfly nymph	Protists

Smallest population

A Food Chain, continued

2. What statement can you make about the amount of energy available at each level of the food chain?

_____

_____

3. There are thousands of individual food chains in the many ecosystems on Earth. Use your encyclopedia and dictionary to find the names of five organisms that live in the same habitat.

A _____

B _____

C _____

D _____

E _____

Smallest population

4. Arrange your five organisms in an order that shows how energy might flow from organism to organism.

Start here

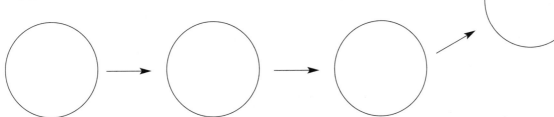

Largest population

5. How is the food chain you created like the pickerel's food chain?

_____

_____

_____

Explore Further

List some populations that could not survive in this habitat. Why could they not survive?

Observing Phototropism

Purpose   How do plants adapt to light? In this investigation, you will observe a plant exhibiting phototropism.

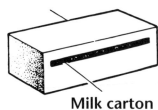

Materials   scissors   milk carton
        potting soil   radish seeds
        shoe box   metric ruler

Procedure

1. Cut a milk carton in half lengthwise. Safety Alert: Always cut in a direction away from your face and body. Plant some radish seeds in the carton and put the plants in normal light until they begin to grow. Answer question 1 below.

2. Turn the shoe box upside down, and make a 0.7 centimeter slot about $\frac{3}{4}$ of the way up.

3. Put the shoe box over the milk carton and arrange it so that light strikes the slot in the shoe box.

4. Leave the plants set up this way for a few days.

Questions and Conclusions

1. Make a hypothesis about what will happen at the end of the experiment. How will the radish plants act in response to light?

   _____

   _____

2. Describe what happened at the conclusion of the experiment. Was your hypothesis supported by the results?

   _____

3. What term is used to describe the plant response you observed?

   _____

4. What advantage might this type of plant response provide plants?

   _____

Explore Further

Repeat this investigation using other seeds, such as nasturtiums, corn, or beans. Do they respond differently?

Comparing Reaction Times

Purpose   How quickly can you catch a falling object?
                  In this investigation, you will compare
                  reaction times.

Materials   meterstick

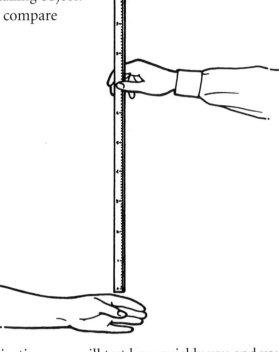

Procedure

1. Look at the figure above. In this investigation, you will test how quickly you and your partner can catch a falling meterstick. *Safety Alert: Be sure the meterstick is not splintered.*

2. Have your partner hold a hand out, as if to grasp something. Hold the meterstick straight up and down above your partner's hand so that the zero mark is just above your partner's hand.

3. Without signaling your partner, let go of the meterstick. Your partner should try to catch the meterstick as quickly as he or she can.

4. After your partner catches the meterstick, record the centimeter mark nearest your partner's thumb. Record this data in the chart below under Try 1.

5. Repeat steps 1–4 four more times. Record the results in the chart below.

6. Switch roles with your partner. Repeat steps 1–5.

	Try 1 (in cm)	Try 2 (in cm)	Try 3 (in cm)	Try 4 (in cm)	Try 5 (in cm)
Your partner's reaction times					
Your reaction times					

Comparing Reaction Times, continued

Questions and Conclusions

1. Compare the results of your partner's first and last tries. Is there a pattern?

   _____

2. Compare the results of your first and last tries. Is there a pattern?

   _____

3. Explain why the results may have changed across the tries.

   _____

   _____

4. Is the behavior you measured in this lab innate or learned? Explain your answer.

   _____

   _____

Explore Further

Repeat this investigation with your partner. This time extend each turn to ten tries. Compare the results to the first set of tries. Explain any differences you observe.

13   Observing Learning Patterns

Use with Investigation 13 on pages 339–340.

Purpose   Can you learn to do a task faster by practicing it? In this investigation, you will compare the learning rates for completing a maze.

Partner's Name	Try 1 (Time in Seconds)	Try 2 (Time in Seconds)	Try 3 (Time in Seconds)	Try 4 (Time in Seconds)	Try 5 (Time in Seconds)

Questions and Conclusions

1. Compare the times for each partner's first try with his or her last try. Do you see a pattern?

_____

_____

2. Explain why the times may have changed across the tries.

_____

_____

Explore Further

1. Have each partner retry your maze five more times. Record the times in a chart. Then graph the times for each partner's tries.

2. Did each time increase, decrease, or stay the same?

14 Making Molds

Use with Investigation 14 on pages 366–367.

Purpose How are fossils formed? In this investigation, you will learn how to make molds of plant and animal remains. This will help you understand the process of fossil formation.

	Observations
Mold of shell	
Mold of leaf	

Questions and Conclusions

1. How are the shell and leaf similar to their molds?

_____

_____

_____

2. How are the shell and leaf different from their molds?

_____

_____

_____

3. Based on your observations in this investigation, what kinds of remains would make the best fossil molds?

_____

_____

_____

Explore Further

Cover each mold with a thin layer of petroleum jelly. Then use more plaster to make casts.

Adaptive Advantage

Purpose   What is the effect of variations, such as different colors, on individuals in a population? In this investigation, you will demonstrate the importance of protective coloration to survival.

Materials   100 green toothpicks
100 red toothpicks
stopwatch or watch with a second hand
green construction paper
red construction paper

Procedure

1. With your partner, decide which of you will be the bird and which will keep the time.

2. Mark off an area that is about 1 meter square. Cover the area with green construction paper.

3. If you are the timekeeper, scatter all 200 of the toothpicks within the area. The toothpicks will represent food for a bird.

4. The partner acting as the bird will place one hand behind his or her back and try to pick up as many toothpicks as possible within a 30-second period. The timekeeper tells the bird when to start and stop picking up toothpicks.

5. At the end of 30 seconds, count the number of toothpicks that were gathered. Record the data in the table on the next page.

6. Repeat steps 3–5 two more times.

7. Repeat steps 3–6 using red construction paper instead of green. Record your data in the table on the next page.

8. Divide the total number of toothpicks picked up for each color by 100 to get the percentage of toothpicks of each color that were picked up.

Adaptive Advantage, continued

Trial	Color of background	Total number of toothpicks		Total number of toothpicks picked up		Percentage of toothpicks picked up	
		Red toothpicks	Green toothpicks	Contrasting background	Matching background	Contrasting background	Matching background
1		100	100				
2		100	100				
3		100	100				
4		100	100				
5		100	100				
6		100	100				

Questions and Conclusions

1. Which color of toothpicks could you see more easily on the green background?

_____

2. Which color of toothpicks could you see more easily on the red background?

_____

3. Compare the percentages of toothpicks picked up. How does the color of toothpicks and the background color relate to the number of each that you collected?

_____

4. Suppose that the toothpicks are insects that a bird eats. How does the color of the insects relate to their chances of survival?

Explore Further

You just investigated the advantage to an organism of camouflage. Color or pattern causes it help it blend into its environment. Some organisms have mimicry. They have a color, shape, or pattern that makes them look like another organism. The other organism is poisonous or tastes bad. Predators have learned to avoid it. Find pictures illustrating these two adaptations. List examples you find and explain why they give the organisms an advantage.

Comparing Body Parts

Purpose   What do similar body parts suggest about the relationship of organisms? In this investigation, you will compare the body parts of four organisms.

Materials   AGS *Biology* textbook

Procedure

1. Turn to page 374 of your textbook. Locate the figure of the penguin, alligator, bat, and human.

2. Compare the front limbs of the four animals in the figure. As you make your observations and comparisons, record your data in the table below.

Animal	Number of bones in upper limb	Number of bones in lower limb	Arrangement of the bones	Function of limb
Penguin				
Alligator				
Bat				
Human				

Questions and Conclusions

1. How are the front limbs of the four animals alike?

   _____

2. How are the front limbs of the four animals different?

   _____

3. Which do you think is probably a better indicator of relationship between two organisms: their structures or the functions of their structures? Explain your answer using data from this investigation.

   _____

   _____

Explore Further

Choose two of the organisms pictured on page 374 in your textbook. Think about how each uses its limbs. Tell why you think each organism's limbs evolved as they did.